Chemistry Games

Volume 1
Chemical Names, Formulas, and Equations

Gregory Gebhart

ISBN-13: 978-1461138990
ISBN-10: 146113899x

Dedication

This book is dedicated to my wife, Fran, mother, Eve, father, Howard, and brothers Brian and Eric. It is also dedicated to my teachers and professors, as well as my supervisors and co-workers. Finally it is dedicated to the one true living God

Come & Go Learn Chemistry with Me - Collect one Element AMU Card Set

Get enough of the elements K and S to correctly make the following chemicals (figure out their formulas from the following names): Potassium (a metal), Octasulfide (a non-metal), and Potassium Sulfide (an ionic compound).

Get enough of the elements Ca and O to correctly make the following chemicals (figure out their formulas from the following names): Calcium (a metal), Dioxygen (a non-metal), Calcium Oxide (an ionic compound).

Get enough of the elements Al and I to correctly make the following chemicals (figure out their formulas from the following names): Aluminum (a metal), Iodine (a non-metal) and Aluminum Iodide (an ionic compound).

Get enough of the elements S, O, and H to correctly make the following chemicals (figure out their formulas from the following names): Sulfur Dioxide (a non-metal oxide), Water (a molecule), and Sulfurous Acid (an oxyacid).

Get enough of the elements Mg, O, and H to correctly make the following chemicals (figure out their formulas from the following names): Magnesium (II) Oxide (a metal oxide), Water (a molecule), and Magnesium (II) Hydroxide (a strong base and an ionic compound).

Get enough of the elements H and O to correctly make the following chemicals (figure out their formulas from the following names): Dihydrogen (a non-metal), Dioxygen (a non-metal) and Water (a molecule).

Get enough of the elements Na, H, and O to correctly make the following chemicals (figure them out from the following names): Sodium Oxide (a metal oxide), Water (a molecule), an Sodium Hydroxide (a metal hydride —a strong base).

Chemical Builder II (Part 1) Game Rules ™ © 2006 GH Gebhart - Reproducible Classroom Document:

(1) Choose a **banker. Select a Game Piece.**

(2) Each player **gets one full set of Element AMU Cards** at **the** start of the game **from** the **banker**. These are all of the elements used in the game (the entire set of elements in the periodic table).

(3) Every time a player **passes the "Come and Go Learn Chemistry with Me" box with his or her Game Piece,** he or she **gets another complete Element AMU set.**

(4) To decide the order of play, each player **selects a Step Card.** The **highest number of** steps on a Step Card **goes first, and so forth**.

(5) On each player's turn, that player must select a Step Card (the deck may be shuffled periodically – no pun intended).

(6) Move one square for each of the steps shown on the Step Card along the square of the Outer Path (**the colored squares**) with his or her **Game Piece.**

(7) After a player has moved, he or she can **decide whether or not to "claim" the square** on the Outer Path he or she landed on (provided no other player already has "claimed" it).

(8) If a player wants **to "claim" the square** on the Outer Path, he or she must **put at least one Element AMU Card** in the Inner Path square (**white square**) **underneath** the Outer Path (colored) square you "claimed." **Get a Claim Card from the banker the corresponding to the Outer Path Square (colored) that you wish to "claim."** The Element AMU Card you put in the Inner Path square must be one of the elements named in the Outer Path square you claimed.

(9) Keep taking turns, moving around the Outer Path squares with your Game Piece after selecting a Step Card.

(10) **If you land on an Outer Path square "claimed" by another player, then** you must **give the other player the same Element AMU Cards as the player has in the Inner Path square underneath it.** If you do not have the same Element AMU Cards, then you must give Element AMU Cards to the player greater than (more than) the AMU (atomic mass unit) of each element whose AMU card you do not have.

(11) Keep taking turns, moving around the Outer Path squares with your Game Piece after selecting a Step Card.

(12) Next, **work on getting all of the elements in each chemical compound's formula named in the Outer Path (colored) square you "claimed."** You must **put one Element AMU Card for each element that you need to correctly write the chemical formula of each named chemical.**

(13) Keep taking turns, moving around the Outer Path squares with your Game Piece after selecting a Step Card.

(14) If you land on another player's Outer Path square, then you must **give to the player "claiming" the square your Element AMU Cards corresponding to the player's Element AMU Cards.** If you do not have the same Element AMU Cards, then you must give Element AMU Cards to the player greater than (more than) the AMU (atomic mass unit) of each element whose Element AMU Card you do not have.

(15) Keep taking turns.

(16) The person with the most Element AMU Cards when "time" is called wins the game.

(17) When a player runs out of AMU cards they must withdraw from the game.

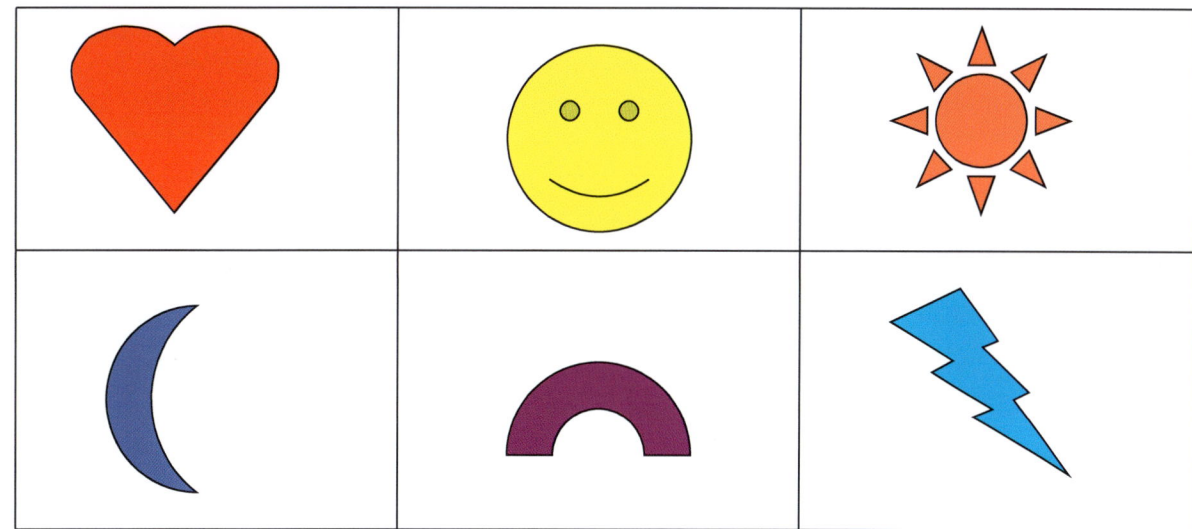

Chemical Builder II (Part 1) Element AMU Cards ™

© 2006 GH Gebhart - Reproducible Classroom Document

In this version of the periodic table of the elements you see the different types of the outermost atomic orbitals (s, p, d, f) that the "valence" electrons of the elements are found in. The negatively charged electrons "orbit" their atoms' small nuclei (where the positively charged protons and electrically neutral neutrons are). This gives the s-block, the p-block, the d-block- and the f-block elemnts.

1 H Hydrogen 1.01 $1s^1$	**2** He Helium 4.00 $1s^2$

3 Li Lithium 6.94 $2s^1$	**4** Be Berylliu m 9.01 $2s^2$
11 Na Sodium 22.99 $3s^1$	**12** Mg Magnesi um 24.31 $3s^2$
19 K Potassiu m 39.10 $4s^1$	**20** Ca Calcium 40.08 $4s^2$
37 Rb Rubidiu m 85.469 $5s^1$	**38** Sr Strontiu m 87.82 $5s^2$
55 Cs Cesium 132.91 $6s^1$	**56** Ba Barium 137.33 $6s^2$
87 Fr Franciu m (223) $7s^1$	**88** Ra Radium 226.03 $7s^2$

p-block

5 B Boron 10.81 $2s^22p^1$	**6** C Carbon 12.01 $2s^22p^2$	**7** N Nitroge n 14.01 $2s^22p^3$	**8** O Oxygen 16.00 $2s^22p^4$	**9** F Fluorine 19.00 $2s^22p^5$	**10** Ne Neon 20.18 $2s^22p^6$
13 Al Alumin um 26.98 $3s^23p^1$	**14** Si Silicon 28.09 $3s^23p^2$	**15** P Phospho rus 30.97 $3s^23p^3$	**16** S Sulfur 32.07 $3s^23p^4$	**17** Cl Chlorine 35.45 $3s^23p^5$	**18** Ar Argon 83.90 $3s^23p^6$
31 Ga Gallium 67.72 $4s^24p^1$	**32** Ge Germani um 72.61 $4s^24p^2$	**33** As Arsenic 74.92 $4s^24p^3$	**34** Se Seleniu m 78.96 $4s^24p^4$	**35** Br Bromine 79.90 $4s^24p^5$	**36** Kr Krypton 83.80 $4s^24p^6$
49 In Indium 114.82 $5s^25p^1$	**50** Sn Tin 116.69 $5s^25p^2$	**51** Sb Antimo ny 121.75 $5s^25p^3$	**52** Te Telluriu m 127.6 $5s^25p^4$	**53** I Iodine 126.9 $5s^25p^5$	**54** Xe Xenon 131.3 $5s^25p^6$
81 Tl Thalliu m 204.37 $6s^26p^1$	**82** Pb Lead 207.2 $6s^26p^2$	**83** Bi Bismuth 208.96 $6s^26p^3$	**84** Po Poloniu m (208) $6s^26p^4$	**85** At Astatine (210) $6s^26p^5$	**86** Rn Radon (222) $6s^26p^6$

d-block

21 Sc Scandiu m 44.96 $4s^23d^1$	**22** Ti Titanium 47.87 $4s^23d^2$	**23** V Vanadiu m 50.94 $4s^23d^3$	**24** Cr Chromiu m 52.00 $4s^13d^5$	**25** Mn Mangane se 54.94 $4s^23d^5$	**26** Fe Iron 55.85 $4s^23d^6$	**27** Co Cobalt 58.93 $4s^23d^7$	**28** Ni Nickel 58.69 $4s^23d^8$	**29** Cu Copper 63.54 $4s^13d^{90}$	**30** Zn Zinc 65.39 $4s^23d^{10}$
39 Y Yttrium 88.908 $5s^24d^1$	**40** Zr Zirconiu m 91.22 $5s^24d^2$	**41** Nb Nobeliu m 92.906 $5s^14d^4$	**42** Mo Molybde num 95.94 $5s^14d^5$	**43** Tc Techneti um (97) $5s^24d^5$	**44** Ru Rutheniu m 102.91 $5s^14d^7$	**45** Rh Rhodiu m 102.91 $5s^14d^8$	**46** Pd Palladiu m 106.4 $4d^{10}$	**47** Ag Silver 107.87 $5s^14d^{10}$	**48** Cd Cadmiu m 112.4 $5s^24d^{10}$
71 Lu Lutetium 174.97 $6s^25d^1$	**72** Hf Hafnium 172.49 $6s^25d^2$	**73** Ta Tantalu m 190.95 $6s^25d^3$	**74** W Tungsto n 183.85 $6s^25d^4$	**75** Re Rhenium 186.21 $6s^25d^5$	**76** Os Osmium 190.2 $6s^25d^6$	**77** Ir Iridium 192.22 $6s^25d^7$	**78** Pt Platinum 195.09 $6s^15d^9$	**79** Au Gold 196.97 $6s^15d^{10}$	**80** Hg Mercury 200.59 $6s^25d^{10}$
103 Lr Lawrenci um (260) $7s^26d^1$	**104** Unq Unnilqu adium (261) $7s^26d^2$	**105** Unp Unnilpe ntium (262) $7s^26d^3$	**106** Unh Unnilhex ium (263) $7s^26d^4$	**107** Uns Unnilsep tium (262) $7s^26d^5$	**108** Uno Unniloct um (265) $7s^26d^6$	**109** Uun Ununiliu m (269) $7s^26d^7$	**110** Uun Ununiliu m (269)	**111** Uuu Ununnun um (272)	**112** Uub

f-block

57 La Lanthan um 138.91 $6s^25d^1$	**58** Ce Cesium 140.12 $6s^25d^14f^1$	**59** Pr Praseody mitum 140.907 $6s^24f^3$	**60** Nd Neodym ium 144.24 $6s^24f^4$	**61** Pm Promethi um (145) $6s^24f^5$	**62** Sm Samariu m 150.36 $6s^24f^6$	**63** Eu Europiu m 151.965 $6s^24f^7$	**64** Gd Gadolini um 157.25 $6s^25d^14f$	**65** Tb Terebiu m 158.93 $6s^24f^9$	**66** Dy Dyspros ium 162.50 $6s^24f^{10}$	**67** Ho Holmiu m 164.93 $6s^24f^{11}$	**68** Er Erbium 167.26 $6s^24f^{12}$	**69** Tm Thulium 168.93 $6s^24f^{13}$	**70** Yb Ytterbiu m 173.04 $6s^24f^{14}$
89 Ac Actiniu m (227) $7s^26d^1$	**90** Th Thallium 232.04 $7s^26d^2$	**91** Pa Protactin ium 231.03 $7s^26d^15f^1$	**92** U Uranium 238.029 $7s^26d^15f^1$	**93** Np Neptuni um (237) $7s^26d^15f^4$	**94** Pu Plutoniu m (244) $7s^25f^6$	**95** Am Americiu m (243) $7s^25f^7$	**96** Cm Curium (247) $7s^26d^15f^7$	**97** Bk Berkeliu m (247) $7s^25f^9$	**98** Cf Californ ium (251) $7s^25f^{10}$	**99** Es Einsteini um (254) $7s^25f^{11}$	**100** Fm Fermiu m (257) $7s^25f^{12}$	**101** Md Mendele vitum (258) $7s^25f^{13}$	**102** No Nobeliu m (259) $7s^25f^{14}$

Move one (1) step. Moving 1 step is to remind you of the element Hydrogen's one proton. Because Hydrogen has one proton, its atomic number is 1. The chemical symbol for the element Hydrogen is H. Because H has one proton and an atomic number = 1, it may be written as $_1H$.	**Move two (2) steps.** Moving 2 steps is to remind you of the element Helium's two protons. Because Helium has two protons, its atomic number is 2. The chemical symbol for the element Helium is He. Because He has two protons and an atomic number = 2, it may be written as $_2He$.	**Move three (3) steps.** Moving 3 steps is to remind you of the element Lithium's three protons. Because Lithium has three protons, its atomic number is 3. The chemical symbol for the element Lithium is Li. Because Li has three protons and an atomic number = 3, it may be written as $_3Li$.	**Move four (4) steps.** Moving 4 steps is to remind you of the element Beryllium's four protons. Because Beryllium has four protons, its atomic number is 4. The chemical symbol for the element Beryllium is Be. Because Be has four protons and an atomic number = 4, it may be written as $_4Be$.
Move five (5) steps. Moving 5 steps is to remind you of the element Boron's five protons. Because Boron has five protons, its atomic number is 5. The chemical symbol for the element Boron is B. Because B has five protons and an number = 5, it may be written as $_5B$.	**Move six (6) steps.** Moving 6 steps is to remind you of the element Carbon's six protons. Because Carbon has six protons, its atomic number is 6. The chemical symbol for the element Carbon is C. Because C has six protons and an atomic number = 6, it may be written as $_6C$.	**Move seven (7) steps.** Moving 7 steps is to remind you of the element Nitrogen's seven protons. Because Nitrogen has seven protons, its atomic number is 7. The chemical symbol for the element Nitrogen is N. Because N has seven protons and an atomic number = 7, it may be written as $_7N$.	**Move eight (8) steps.** Moving 8 steps is to remind you of the element Oxygen's eight protons. Because Oxygen has eight protons, its atomic number is 8. The chemical symbol for the element Oxygen is O. Because O has eight protons and an atomic number = 8, it may be written as $_8O$.

Chemical Builder II (Part 1) Claim Cards ™ © 2006 GH Gebhart - Reproducible Classroom Document

Get enough of the elements K and S to correctly make the following chemicals (figure out their formulas from the following names): Potassium (a metal), Octasulfide (a non-metal), and Potassium Sulfide (an ionic compound). Answer: K, S_8, K_2S

Get enough of the elements Ca and O to correctly make the following chemicals (figure out their formulas from the following names): Calcium (a metal), Dioxygen (a non-metal), Calcium Oxide (an ionic compound). Answer: Ca, O_2, CaO

Get enough of the elements Al and I to correctly make the following chemicals (figure out their formulas from the following names): Aluminum (a metal), Iodine (a non-metal) and Aluminum Iodide (an ionic compound). Answer: Al, I_2, AlI_3

Get enough of the elements S, O, and H to correctly make the following chemicals (figure out their formulas from the following names): Sulfur Dioxide (a non-metal oxide), Water (a molecule), and Sulfurous Acid (an oxyacid). Answer: SO_2, H_2O, H_2SO_3

Get enough of the elements H and O to correctly make the following chemicals (figure out their formulas from the following names): Dihydrogen (a non-metal), Dioxygen (a non-metal) and Water (a molecule). Answer: H_2, O_2, H_2O

Get enough of the elements Mg, O, and H to correctly make the following chemicals (figure out their formulas from the following names): Magnesium (II) Oxide (a metal oxide), Water (a molecule), and Magnesium (II) Hydroxide (a strong base and an ionic compound). Answer: MgO, H_2O, $Mg(OH)_2$

Get enough of the elements Na, H, and O to correctly make the following chemicals (figure them out from the following names): Sodium Oxide (a metal oxide), Water (a molecule), an Sodium Hydroxide (a metal hydride –a strong base). Answer: Na_2O, H_2O, NaOH

Come & Go Learn Chemistry with Me - Collect one Element AMU Card Set

Get enough of the elements Fe, O, and H to correctly make the following chemical compounds (figure out their formulas from the following names):
Iron (III) Hydroxide (a metal hydroxide), Iron (III) Oxide (a metal oxide), and Water (a molecule).

Get enough of the elements Ca, O, and S to correctly make the following chemicals (figure out their formulas from the following names):
Calcium Oxide (a metal oxide), Sulfur Trioxide (a non-metal oxide), and Calcium Sulfate (an Oxysalt) - only if water is absent.

Get enough of the elements Na, O, H, and C to correctly make the following chemicals (figure out their formulas from the following names): Sodium Hydroxide (a metal oxide), Carbon Dioxide (a non-metal oxide), and Sodium Bicarbonate (a Oxysalt) - only if water is not present.

Get enough of the elements N, H, N, O, and Ca to correctly make the following chemicals (figure out their formulas from the following names): Ammonium Nitrate (an ionic compound), Calcium Oxide (a metal oxide), Ammonia, Water, and Calcium Nitrate..

Get enough of the elements A, O, H, Cl to correctly make the following chemicals (figure out their formulas from the following names): Aluminum Oxide (a weak base), Hydrochlorous Acid (a weak acid), Aluminum Chlorate (a metal salt), and Water (a molecule).

Get enough of the elements H, Cl, Ca, O, and H to correctly make the following chemicals (figure out their formulas from the following names): Hydrochloric Acid (a strong acid), Calcium Hydroxide (a strong base), Calcium Chloride (a metal salt), Water (a molecule).

Get enough of the elements H, S, O, and Fe to correctly make the following chemicals (figure them out from the following names): Sulfuric Acid (a strong acid), Iron (III) Hydroxide (a strong base), Iron (III) Sulfate (a metal salt), and Water (a molecule).

(18)   **Choose** a banker. **Select a Game Piece.**

(19)   Each player **gets one full set of Element AMU Cards** at **the** start of the game **from** the **banker.** These are all of the elements used in the game (the entire set of elements in the periodic table).

(20)   Every time a player **passes the "Come and Go Learn Chemistry with Me" box with his or her Game Piece,** he or she **gets another complete Element AMU set.**

(21)   **To decide the order of play,** each player **selects a Step Card.** The **highest number** of steps on a Step Card **goes first, and so forth.**

(22)   **On each player's turn, that player must select a Step Card** (the deck may be shuffled periodically – no pun intended).

(23)   **Move one square for each of the steps shown on the Step Card along** the square of the Outer Path (**the colored squares**) with his or her **Game Piece.**

(24)   After a player has moved, he or she can **decide whether or not to "claim" the square** on the Outer Path he or she landed on (provided no other player already has "claimed" it).

(25)   If a player wants **to "claim" the square** on the Outer Path, he or she must **put at least one Element AMU Card** in the Inner Path square (**white square**) **underneath** the Outer Path (colored) square you "claimed." Get a Claim Card **from the banker the corresponding to the Outer Path Square (colored) that you wish to "claim."** The Element AMU Card you put in the Inner Path square must be one of the elements named in the Outer Path square you claimed.

(26)   **Keep taking turns, moving around the Outer Path squares with your Game Piece after selecting a Step Card.**

(27)   **If you land on an Outer Path square "claimed" by another player, then** you must **give the other player the same Element AMU Cards as the player has in the Inner Path square underneath it.** If you do not have the same Element AMU Cards, then you must give Element AMU Cards to the player greater than (more than) the AMU (atomic mass unit) of each element whose AMU card you do not have.

(28)   **Keep taking turns, moving around the Outer Path squares with your Game Piece after selecting a Step Card.**

(29)   **Next, work on getting all of the elements in each chemical compound's formula named in the Outer Path (colored) square you "claimed."** You must **put one Element AMU Card for each element that you need to correctly write the chemical formula of each named chemical.**

(30)   **Keep taking turns, moving around the Outer Path squares** with your Game Piece after selecting a Step Card.

(31)   **If you land on another player's Outer Path square, then** you must **give to the player "claiming" the square your Element AMU Cards corresponding to the player's Element AMU Cards.** If you do not have the same Element AMU Cards, then you must give Element AMU Cards to the player greater than (more than) the AMU (atomic mass unit) of each element whose Element AMU Card you do not have.

(32)   **Keep taking turns.**

(33)   **The person with the most Element AMU Cards when "time" is called wins the game.**

(34)   **When a player runs out of AMU cards they must withdraw from the game.**

Chemical Builder II (Part 2) Element AMU Cards ™

© 2006 GH Gebhart - Reproducible Classroom Document

In this version of the periodic table of the elements you see the different types of the outermost atomic orbitals (s, p, d, f) that the "valence" electrons of the elements are found in. The negatively charged electrons "orbit" their atoms' small nuclei (where the positively charged protons and electrically neutral neutrons are). This gives the s-block, the p-block, the d-block, and the f-block elemnts.

1 H Hydrogen 1.01 $1s^1$		2 He Helium 4.00 $1s^2$

3 Li Lithium 6.94 $2s^1$	4 Be Beryllium 9.01 $2s^2$

5 B Boron 10.81 $2s^22p^1$	6 C Carbon 12.01 $2s^22p^2$	7 N Nitrogen 14.01 $2s^22p^3$	8 O Oxygen 16.00 $2s^22p^4$	9 F Fluorine 19.00 $2s^22p^5$	10 Ne Neon 20.18 $2s^22p^6$

11 Na Sodium 22.99 $3s^1$	12 Mg Magnesium 24.31 $3s^2$

13 Al Aluminum 26.98 $3s^23p^1$	14 Si Silicon 28.09 $3s^23p^2$	15 P Phosphorus 30.97 $3s^23p^3$	16 S Sulfur 32.07 $3s^23p^4$	17 Cl Chlorine 35.45 $3s^23p^5$	18 Ar Argon 83.90 $3s^23p^6$

19 K Potassium 39.10 $4s^1$	20 Ca Calcium 40.08 $4s^2$	21 Sc Scandium 44.96 $4s^23d^1$	22 Ti Titanium 47.87 $4s^23d^2$	23 V Vanadium 50.94 $4s^23d^3$	24 Cr Chromium 52.00 $4s^13d^5$	25 Mn Manganese 54.94 $4s^23d^5$	26 Fe Iron 55.85 $4s^23d^6$	27 Co Cobalt 58.93 $4s^23d^7$	28 Ni Nickel 58.69 $4s^23d^8$	29 Cu Copper 63.54 $4s^13d^{90}$	30 Zn Zinc 65.39 $4s^23d^{10}$	31 Ga Gallium 67.72 $4s^24p^1$	32 Ge Germanium 72.61 $4s^24p^2$	33 As Arsenic 74.92 $4s^24p^3$	34 Se Selenium 78.96 $4s^24p^4$	35 Br Bromine 79.90 $4s^24p^5$	36 Kr Krypton 83.80 $4s^24p^6$

37 Rb Rubidium 85.469 $5s^1$	38 Sr Strontium 87.82 $5s^2$	39 Y Yttrium 88.908 $5s^24d^1$	40 Zr Zirconium 91.22 $5s^24d^2$	41 Nb Nobelium 92.906 $5s^14d^4$	42 Mo Molybdenum 95.94 $5s^24d^4$	43 Tc Technetium (97) $5s^24d^5$	44 Ru Ruthenium 102.91 $5s^14d^7$	45 Rh Rhodium 102.91 $5s^14d^8$	46 Pd Palladium '106.4 $4d^{10}$	47 Ag Silver 107.87 $5s^14d^{10}$	48 Cd Cadmium 112.4 $5s^24d^{10}$	49 In Indium 114.82 $5s^25p^1$	50 Sn Tin 116.69 $5s^25p^2$	51 Sb Antimony 121.75 $5s^25p^3$	52 Te Tellurium 127.6 $5s^25p^4$	53 I Iodine 126.9 $5s^25p^5$	54 Xe Xenon 131.3 $5s^25p^6$

55 Cs Cesium 132.91 $6s^1$	56 Ba Barium 137.33 $6s^2$	71 Lu Lutetium 174.97 $6s^25d^1$	72 Hf Hafnium 172.49 $6s^25d^2$	73 Ta Tantalum 190.95 $6s^25d^3$	74 W Tungsten 183.85 $6s^25d^4$	75 Re Rhenium 186.21 $6s^25d^5$	76 Os Osmium 190.2 $6s^25d^6$	77 Ir Iridium 192.22 $6s^25d^7$	78 Pt Platinum 195.09 $6s^15d^9$	79 Au Gold 196.97 $6s^15d^{10}$	80 Hg Mercury 200.59 $6s^25d^{10}$	81 Tl Thallium 204.37 $6s^26p^1$	82 Pb Lead 207.2 $6s^26p^2$	83 Bi Bismuth 208.96 $6s^26p^3$	84 Po Polonium (208) $6s^26p^4$	85 At Astatine (210) $6s^26p^5$	86 Rn Radon (222) $6s^26p^6$

87 Fr Francium (223) $7s^1$	88 Ra Radium 226.03 $7s^2$	103 Lr Lawrencium (260) $7s^26d^1$	104 Unq Unnilquadium (261) $7s^26d^2$	105 Unp Unnilpentium (262) $7s^26d^3$	106 Unh Unnilhexium (263) $7s^26d^4$	107 Uns Unnilseptium (262) $7s^26d^5$	108 Uno Unniloctium (265) $7s^26d^6$	109 Une Ununilnium (269) $7s^26d^7$	110 Uun Ununilium (272)	111 Uuu Unununium (272)	112 Uub

57 La Lanthanum 138.91 $6s^25d^1$	58 Ce Cesium 140.12 $6s^25d^14f^1$	59 Pr Praseodymium 140.907 $6s^24f^3$	60 Nd Neodymium 144.24 $6s^24f^4$	61 Pm Promethium (145) $6s^24f^5$	62 Sm Samarium 150.36 $6s^24f^6$	63 Eu Europium 151.965 $6s^24f^7$	64 Gd Gadolinium 157.25 $6s^25d^14f^7$	65 Tb Terebium 158.93 $6s^24f^9$	66 Dy Dysprosium 162.50 $6s^24f^{10}$	67 Ho Holmium 164.93 $6s^24f^{11}$	68 Er Erbium 167.26 $6s^24f^{12}$	69 Tm Thulium 168.93 $6s^24f^{13}$	70 Yb Ytterbium 173.04 $6s^24f^{14}$

89 Ac Actinium (227) $7s^26d^1$	90 Th Thallium 232.04 $7s^26d^2$	91 Pa Protactinium 231.03 $7s^26d^15f^2$	92 U Uranium 238.029 $7s^26d^15f^3$	93 Np Neptunium (237) $7s^26d^15f^4$	94 Pu Plutonium (244) $7s^25f^6$	95 Am Americium (243) $7s^25f^7$	96 Cm Curium (247) $7s^26d^15f^7$	97 Bk Berkelium (247) $7s^25f^9$	98 Cf Californium (251) $7s^25f^{10}$	99 Es Einsteinium (254) $7s^25f^{11}$	100 Fm Fermium (257) $7s^25f^{12}$	101 Md Mendelevium (258) $7s^25f^{13}$	102 No Nobelium (259) $7s^25f^{14}$

Move one (1) step. Moving 1 step is to remind you of the element Hydrogen's one proton. Because Hydrogen has one proton, its atomic number is 1. The chemical symbol for the element Hydrogen is H. Because H has one proton and an atomic number = 1, it may be written as $_1H$.	**Move two (2) steps.** Moving 2 steps is to remind you of the element Helium's two protons. Because Helium has two protons, its atomic number is 2. The chemical symbol for the element Helium is He. Because He has two protons and an atomic number = 2, it may be written as $_2He$.	**Move three (3) steps.** Moving 3 steps is to remind you of the element Lithium's three protons. Because Lithium has three protons, its atomic number is 3. The chemical symbol for the element Lithium is Li. Because Li has three protons and an atomic number = 3, it may be written as $_3Li$.	**Move four (4) steps.** Moving 4 steps is to remind you of the element Beryllium's four protons. Because Beryllium has four protons, its atomic number is 4. The chemical symbol for the element Beryllium is Be. Because Be has four protons and an atomic number = 4, it may be written as $_4Be$.
Move five (5) steps. Moving 5 steps is to remind you of the element Boron's five protons. Because Boron has five protons, its atomic number is 5. The chemical symbol for the element Boron is B. Because B has five protons and an number = 5, it may be written as $_5B$.	**Move six (6) steps.** Moving 6 steps is to remind you of the element Carbon's six protons. Because Carbon has six protons, its atomic number is 6. The chemical symbol for the element Carbon is C. Because C has six protons and an atomic number = 6, it may be written as $_6C$.	**Move seven (7) steps.** Moving 7 steps is to remind you of the element Nitrogen's seven protons. Because Nitrogen has seven protons, its atomic number is 7. The chemical symbol for the element Nitrogen is N. Because N has seven protons and an atomic number = 7, it may be written as $_7N$.	**Move eight (8) steps.** Moving 8 steps is to remind you of the element Oxygen's eight protons. Because Oxygen has eight protons, its atomic number is 8. The chemical symbol for the element Oxygen is O. Because O has eight protons and an atomic number = 8, it may be written as $_8O$.

Chemical Builder II (Part 2) Claim Cards (Answers too) ™ © 2006 GH Gebhart - Reproducible

Get enough of the elements Fe, O, and H to correctly make the following chemical compounds (figure out their formulas from the following names): Iron (III) Hydroxide (a metal hydroxide), Iron (III) Oxide (a metal oxide), and Water (a molecule). Answer: $Fe(OH)_3$, Fe_2O_3, H_2O

Get enough of the elements Ca, O, and S to correctly make the following chemicals (figure out their formulas from the following names): Calcium Oxide (a metal oxide), Sulfur Trioxide (a non-metal oxide), and Calcium Sulfate (an Oxysalt) - only if water is absent. Answer: $CaO, SO_3, CaSO_4$

Get enough of the elements Na, O, H, and C to correctly make the following chemicals (figure out their formulas from the following names): Sodium Hydroxide (a metal oxide), Carbon Dioxide (a non-metal oxide), and Sodium Bicarbonate (a Oxysalt) - only if water is not present. Answer: $NaOH$, CO_2, $NaHCO_3$

Get enough of the elements N, H, N, O, and Ca to correctly make the following chemicals (figure out their formulas from the following names): Ammonium Nitrate (an ionic compound), Calcium Oxide (a metal oxide), Ammonia, Water, and Calcium Nitrate. Answer: NH_4NO_3, CaO, NH_3, H_2O, $Ca(NO_3)_2$

Get enough of the elements H, Cl, Ca, O, and H to correctly make the following chemicals (figure out their formulas from the following names): Hydrochloric Acid (a strong acid), Calcium Hydroxide (a strong base), Calcium Chloride (a metal salt), Water (a molecule). Answer: HCl, $Ca(OH)_2$, $CaCl_2$, H_2O

Get enough of the elements Al, O, H, Cl to correctly make the following chemicals (figure out their formulas from the following names): Aluminum Oxide (a weak base), Hydrochlorous Acid (a weak acid), Aluminum Chlorate (a metal salt), and Water (a molecule). Answer: Al_2O_3, $HClO$, $Al(ClO_4)_3$, H_2O

Get enough of the elements H, S, O, and Fe to correctly make the following chemicals (figure them out from the following names): Sulfuric Acid (a strong acid), Iron (III) Hydroxide (a strong base), Iron (III) Sulfate (a metal salt), and Water (a molecule). Answer: H_2SO_4, $Fe(OH)_3$, $Fe_2(SO_4)_3$, H_2O

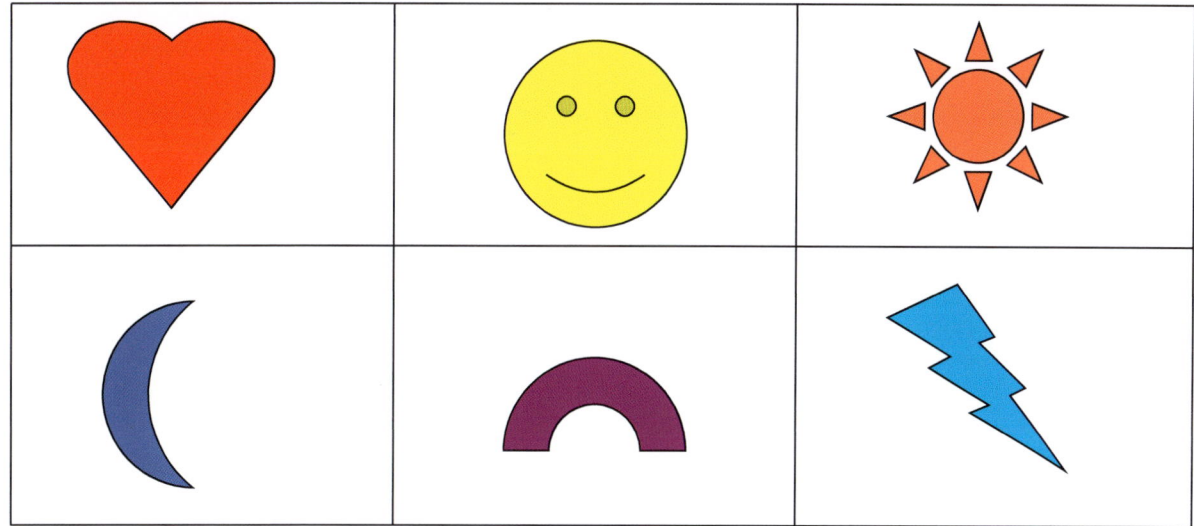

Chemical Builder II (Part 3) Game ™ © 2006 GH Gebhart - Reproducible Classroom Document

Come & Go Learn Chemistry with Me - Collect one Element AMU Card Set	Get enough of the elements Na, H, C, O, and S to correctly make the following chemical compounds (figure out their formulas from the names): Sodium Bicarbonate (salt of a weak acid), Sulfuric Acid (strong acid), Sodium Sulfate (oxyacid), Carbon Dioxide (molecule), and Water.	Get enough of the elements Ba, C, O, H, and Br to correctly make the following chemicals (figure out their formulas from the following names): Barium Carbonate (salt of a weak acid), Hydrobromic Acid (weak acid), Barium DiBromide (salt), Carbon Dioxide (molecule) and Water.	Get enough of the elements Mg, S, H, and Cl to correctly make the following chemicals (figure out their formulas from the following names): Magnesium Sulfide (salt of a weak acid), Hydrochloric Acid (strong acid), Hydrogen Sulfide (weak acid), and Magnesium Chloride (salt).
Get enough of the elements K, Cl, and O to correctly make the following chemicals (figure out their formulas from the following names): Potassium Chlorate (an oxysalt), Potassium Chloride (a metal salt), and Oxygen (a molecule).			Get enough of the elements K, S, O, H, and N to correctly make the following chemicals (figure out their formulas from the following names): Potassium Sulfite (salt of a strong acid), Nitric Acid (strong acid), Potassium Nitrate (weak acid), Sulfur Dioxide (molecule), and Water.
Get enough of the elements Ag, N, O, Fe, and Cl to correctly make the following chemicals (figure out their formulas from the following names): Silver (I) Nitrate, Iron (III) Chloride, Silver (I) Chloride, Iron (III) Nitrate (two soluble salts, a metal salt, and a precipitate).			Get enough of the elements Ca, Cl, K, C, and O to correctly make the following chemicals (figure them out from the following names): Calcium Chloride, Potassium Carbonate, Potassium Chloride, Calcium Carbonate (three soluble salts, and a precipitate).

(35)   **Choose** a **banker**. **Select a Game Piece.**

(36)   Each player **gets one full set of Element AMU Cards** at **the** start of the game **from** the **banker**. These are all of the elements used in the game (the entire set of elements in the periodic table).

(37)   Every time a player **passes the "Come and Go Learn Chemistry with Me" box with his or her Game Piece**, he or she **gets another complete Element AMU set.**

(38)   **To decide the order of play**, each player **selects a Step Card**. The **highest number** of steps on a Step Card **goes first, and so forth.**

(39)   **On each player's turn, that player must select a Step Card** (the deck may be shuffled periodically – no pun intended).

(40)   **Move one square for each of the steps shown on the Step Card along** the square of the Outer Path (**the colored squares**) with his or her **Game Piece.**

(41)   After a player has moved, he or she can **decide whether or not to "claim" the square** on the Outer Path he or she landed on (provided no other player already has "claimed" it).

(42)   If a player wants **to "claim" the square** on the Outer Path, he or she must **put at least one Element AMU Card** in the Inner Path square (**white square**) **underneath** the Outer Path (**colored) square you "claimed." Get a Claim Card from the banker the corresponding to the Outer Path Square (colored) that you wish to "claim."** The Element AMU Card you put in the Inner Path square must be one of the elements named in the Outer Path square you claimed.

(43)   **Keep taking turns, moving around the Outer Path squares with your Game Piece after selecting a Step Card.**

(44)   **If you land on an Outer Path square "claimed" by another player, then** you must **give the other player the same Element AMU Cards as the player has in the Inner Path square underneath it**. If you do not have the same Element AMU Cards, then you must give Element AMU Cards to the player greater than (more than) the AMU (atomic mass unit) of each element whose AMU card you do not have.

(45)   **Keep taking turns, moving around the Outer Path squares with your Game Piece after selecting a Step Card.**

(46)   **Next, work on getting all of the elements in each chemical compound's formula named in the Outer Path (colored) square you "claimed."** You must **put one Element AMU Card for each element that you need to correctly write the chemical formula of each named chemical.**

(47)   Keep taking turns, moving around the Outer Path squares with your Game Piece after selecting a Step Card.

(48)   If you land on another player's Outer Path square, then you must **give to the player "claiming" the square your Element AMU Cards corresponding to the player's Element AMU Cards.** If you do not have the same Element AMU Cards, then you must give Element AMU Cards to the player greater than (more than) the AMU (atomic mass unit) of each element whose Element AMU Card you do not have.

(49)   **Keep taking turns.**

(50)   **The person with the most Element AMU Cards when "time" is called wins the game.**

(51)   **When a player runs out of AMU cards they must withdraw from the game.**

Chemical Builder II (Part 3) Element AMU Cards ™

© 2006 GH Gebhart - Reproducible Classroom Document

In this version of the periodic table of the elements you see the different types of the outermost atomic orbitals (s, p, d, f) that the "valence" electrons of the elements are found in. The negatively charged electrons "orbit" their atoms' small nuclei (where the positively charged protons and electrically neutral neutrons are). This gives the s-block, the p-block, the d-block, and the f-block elemnts.

s-block

- **H** 1 — Hydrogen, 1.01, $1s^1$
- **He** 2 — Helium, 4.00, $1s^2$
- **Li** 3 — Lithium, 6.94, $2s^1$
- **Be** 4 — Beryllium, 9.01, $2s^2$
- **Na** 11 — Sodium, 22.99, $3s^1$
- **Mg** 12 — Magnesium, 24.31, $3s^2$
- **K** 19 — Potassium, 39.10, $4s^1$
- **Ca** 20 — Calcium, 40.08, $4s^2$
- **Rb** 37 — Rubidium, 85.469, $5s^1$
- **Sr** 38 — Strontium, 87.82, $5s^2$
- **Cs** 55 — Cesium, 132.91, $6s^1$
- **Ba** 56 — Barium, 137.33, $6s^2$
- **Fr** 87 — Francium, (223), $7s^1$
- **Ra** 88 — Radium, 226.03, $7s^2$

p-block

- **B** 5 — Boron, 10.81, $2s^22p^1$
- **C** 6 — Carbon, 12.01, $2s^22p^2$
- **N** 7 — Nitrogen, 14.01, $2s^22p^3$
- **O** 8 — Oxygen, 16.00, $2s^22p^4$
- **F** 9 — Fluorine, 19.00, $2s^22p^5$
- **Ne** 10 — Neon, 20.18, $2s^22p^6$
- **Al** 13 — Aluminum, 26.98, $3s^23p^1$
- **Si** 14 — Silicon, 28.09, $3s^23p^2$
- **P** 15 — Phosphorus, 30.97, $3s^23p^3$
- **S** 16 — Sulfur, 32.07, $3s^23p^4$
- **Cl** 17 — Chlorine, 35.45, $3s^23p^5$
- **Ar** 18 — Argon, 83.90, $3s^23p^6$
- **Ga** 31 — Gallium, 67.72, $4s^24p^1$
- **Ge** 32 — Germanium, 72.61, $4s^24p^2$
- **As** 33 — Arsenic, 74.92, $4s^24p^3$
- **Se** 34 — Selenium, 78.96, $4s^24p^4$
- **Br** 35 — Bromine, 79.90, $4s^24p^5$
- **Kr** 36 — Krypton, 83.80, $4s^24p^6$
- **In** 49 — Indium, 114.82, $5s^25p^1$
- **Sn** 50 — Tin, 116.69, $5s^25p^2$
- **Sb** 51 — Antimony, 121.75, $5s^25p^3$
- **Te** 52 — Tellurium, 127.6, $5s^25p^4$
- **I** 53 — Iodine, 126.9, $5s^25p^5$
- **Xe** 54 — Xenon, 131.3, $5s^25p^6$
- **Tl** 81 — Thallium, 204.37, $6s^26p^1$
- **Pb** 82 — Lead, 207.2, $6s^26p^2$
- **Bi** 83 — Bismuth, 208.96, $6s^26p^3$
- **Po** 84 — Polonium, (208), $6s^26p^4$
- **At** 85 — Astatine, (210), $6s^26p^5$
- **Rn** 86 — Radon, (222), $6s^26p^6$

d-block

- **Sc** 21 — Scandium, 44.96, $4s^23d^1$
- **Ti** 22 — Titanium, 47.90, $4s^23d^2$
- **V** 23 — Vanadium, 50.94, $4s^23d^3$
- **Cr** 24 — Chromium, 52.00, $4s^13d^5$
- **Mn** 25 — Manganese, 54.94, $4s^23d^5$
- **Fe** 26 — Iron, 55.85, $4s^23d^6$
- **Co** 27 — Cobalt, 58.93, $4s^23d^7$
- **Ni** 28 — Nickel, 58.69, $4s^23d^8$
- **Cu** 29 — Copper, 63.54, $4s^13d^{10}$
- **Zn** 30 — Zinc, 65.39, $4s^23d^{10}$
- **Y** 39 — Yttrium, 88.908, $5s^24d^1$
- **Zr** 40 — Zirconium, 91.22, $5s^24d^2$
- **Nb** 41 — Nobelium, 92.906, $5s^14d^4$
- **Mo** 42 — Molybdenum, 95.94, $5s^24d^4$
- **Tc** 43 — Technetium, (97), $5s^24d^5$
- **Ru** 44 — Ruthenium, 102.91, $5s^14d^7$
- **Rh** 45 — Rhodium, 102.91, $5s^14d^8$
- **Pd** 46 — Palladium, 106.4, $4d^{10}$
- **Ag** 47 — Silver, 107.87, $5s^14d^{10}$
- **Cd** 48 — Cadmium, 112.4, $5s^24d^{10}$
- **Lu** 71 — Lutetium, 174.97, $6s^25d^1$
- **Hf** 72 — Hafnium, 172.49, $6s^25d^2$
- **Ta** 73 — Tantalum, 190.95, $6s^25d^3$
- **W** 74 — Tungsten, 183.85, $6s^25d^4$
- **Re** 75 — Rhenium, 186.21, $6s^25d^5$
- **Os** 76 — Osmium, 190.2, $6s^25d^6$
- **Ir** 77 — Iridium, 192.22, $6s^25d^7$
- **Pt** 78 — Platinum, 195.09, $6s^15d^9$
- **Au** 79 — Gold, 196.97, $6s^15d^{10}$
- **Hg** 80 — Mercury, 200.59, $6s^25d^{10}$
- **Lr** 103 — Lawrencium, (260), $7s^26d^1$
- **Unq** 104 — Unnilquadium, (261), $7s^26d^2$
- **Unp** 105 — Unnilpentium, (262), $7s^26d^3$
- **Unh** 106 — Unnilhexium, (263), $7s^26d^4$
- **Uns** 107 — Unnilseptium, (262), $7s^26d^5$
- **Uno** 108 — Unniloctium, (265), $7s^26d^6$
- **Une** 109 — Unnilennium, (266), $7s^26d^7$
- **Uun** 110 — Ununilium, (269), $7s^26d^8$
- **Uuu** 111 — Unununium, (272)
- **Uub** 112

f-block

- **La** 57 — Lanthanum, 138.91, $6s^25d^1$
- **Ce** 58 — Cesium, 140.12, $6s^25d^14f^1$
- **Pr** 59 — Praseodymium, 140.907, $6s^24f^3$
- **Nd** 60 — Neodymium, 144.24, $6s^24f^4$
- **Pm** 61 — Promethium, (145), $6s^24f^5$
- **Sm** 62 — Samarium, 150.36, $6s^24f^6$
- **Eu** 63 — Europium, 151.965, $6s^24f^7$
- **Gd** 64 — Gadolinium, 157.25, $6s^25d^14f^7$
- **Tb** 65 — Terebium, 158.93, $6s^24f^9$
- **Dy** 66 — Dysprosium, 162.50, $6s^24f^{10}$
- **Ho** 67 — Holmium, 164.93, $6s^24f^{11}$
- **Er** 68 — Erbium, 167.26, $6s^24f^{12}$
- **Tm** 69 — Thulium, 168.93, $6s^24f^{13}$
- **Yb** 70 — Ytterbium, 173.04, $6s^24f^{14}$
- **Ac** 89 — Actinium, (227), $7s^26d^1$
- **Th** 90 — Thallium, 232.04, $7s^26d^2$
- **Pa** 91 — Protactinium, 231.03, $7s^26d^15f^2$
- **U** 92 — Uranium, 238.029, $7s^26d^15f^3$
- **Np** 93 — Neptunium, (237), $7s^26d^15f^4$
- **Pu** 94 — Plutonium, (244), $7s^25f^6$
- **Am** 95 — Americium, (243), $7s^25f^7$
- **Cm** 96 — Curium, (247), $7s^26d^15f^7$
- **Bk** 97 — Berkelium, (247), $7s^25f^9$
- **Cf** 98 — Californium, (251), $7s^25f^{10}$
- **Es** 99 — Einsteinium, (254), $7s^25f^{11}$
- **Fm** 100 — Fermium, (257), $7s^25f^{12}$
- **Md** 101 — Mendelevium, (258), $7s^25f^{13}$
- **No** 102 — Nobelium, (259), $7s^25f^{14}$

Move one (1) step.	Move two (2) steps.	Move three (3) steps.	Move four (4) steps.
Moving 1 step is to remind you of the element Hydrogen's one proton. Because Hydrogen has one proton, its atomic number is 1. The chemical symbol for the element Hydrogen is H. Because H has one proton and an atomic number = 1, it may be written as $_1H$.	Moving 2 steps is to remind you of the element Helium's two protons. Because Helium has two protons, its atomic number is 2. The chemical symbol for the element Helium is He. Because He has two protons and an atomic number = 2, it may be written as $_2He$.	Moving 3 steps is to remind you of the element Lithium's three protons. Because Lithium has three protons, its atomic number is 3. The chemical symbol for the element Lithium is Li. Because Li has three protons and an atomic number = 3, it may be written as $_3Li$.	Moving 4 steps is to remind you of the element Beryllium's four protons. Because Beryllium has four protons, its atomic number is 4. The chemical symbol for the element Beryllium is Be. Because Be has four protons and an atomic number = 4, it may be written as $_4Be$.
Move five (5) steps.	Move six (6) steps.	Move seven (7) steps.	Move eight (8) steps.
Moving 5 steps is to remind you of the element Boron's five protons. Because Boron has five protons, its atomic number is 5. The chemical symbol for the element Boron is B. Because B has five protons and an number = 5, it may be written as $_5B$.	Moving 6 steps is to remind you of the element Carbon's six protons. Because Carbon has six protons, its atomic number is 6. The chemical symbol for the element Carbon is C. Because C has six protons and an atomic number = 6, it may be written as $_6C$.	Moving 7 steps is to remind you of the element Nitrogen's seven protons. Because Nitrogen has seven protons, its atomic number is 7. The chemical symbol for the element Nitrogen is N. Because N has seven protons and an atomic number = 7, it may be written as $_7N$.	Moving 8 steps is to remind you of the element Oxygen's eight protons. Because Oxygen has eight protons, its atomic number is 8. The chemical symbol for the element Oxygen is O. Because O has eight protons and an atomic number = 8, it may be written as $_8O$.

Get enough of the elements Na, H, C, O, and S to correctly make the following chemical compounds (figure out their formulas from the names): Sodium Bicarbonate (salt of a weak acid), Sulfuric Acid (strong acid), Sodium Sulfate (oxyacid), Carbon Dioxide (molecule), and Water. Answer: $NaHCO_3$, H_2SO_4, Na_2SO_4, CO_2, H_2O

Get enough of the elements Ba, C, O, H, and Br to correctly make the following chemicals (figure out their formulas from the following names): Barium Carbonate (salt of a weak acid), Hydrobromic Acid (weak acid), Barium DiBromide (salt), Carbon Dioxide (molecule) and Water. Answer: $BaCO_3$, HBr, $BaBr_2$, CO_2, H_2O

Get enough of the elements Mg, S, H, and Cl to correctly make the following chemicals (figure out their formulas from the following names): Magnesium Sulfide (salt of a weak acid), Hydrochloric Acid (strong acid), Hydrogen Sulfide (weak acid), and Magnesium Chloride (salt). Answer: MgS, HCl, H_2S, $MgCl_2$

Get enough of the elements K, Cl, and O to correctly make the following chemicals (figure out their formulas from the following names): Potassium Chlorate (an oxysalt), Potassium Chloride (a metal salt), and Oxygen (a molecule). Answer: $KClO_3$, KCl, O_2

Get enough of the elements K, S, O, H, and N to correctly make the following chemicals (figure out their formulas from the following names): Potassium Sulfite (salt of a weak acid), Nitric Acid (strong acid), Potassium Nitrate (weak acid), Sulfur Dioxide (molecule), and Water. Answer: K_2SO_3, HNO_3, KNO_3, SO_2, H_2O

Get enough of the elements Ca, Cl, K, C, and O to correctly make the following chemicals (figure them out from the following names): Calcium Chloride, Potassium Carbonate, Potassium Chloride, Calcium Carbonate (three soluble salts, and a precipitate). Answer: $CaCl_2$, K_2CO_3, KCl, $CaCO_3$.

Get enough of the elements Ag, N, O, Fe, and Cl to correctly make the following chemicals (figure out their formulas from the following names): Silver (I) Nitrate, Iron (III) Chloride, Silver (I) Chloride, Iron (III) Nitrate (two soluble salts, a metal salt, and a precipitate). Answer: $FeCl_3$, $AgNO_3$, AgCl, $Fe(NO_3)_3$

Chemical Builder II (Part 3) Game Pieces ™ © 2006 GH Gebhart - Reproducible Classroom Document

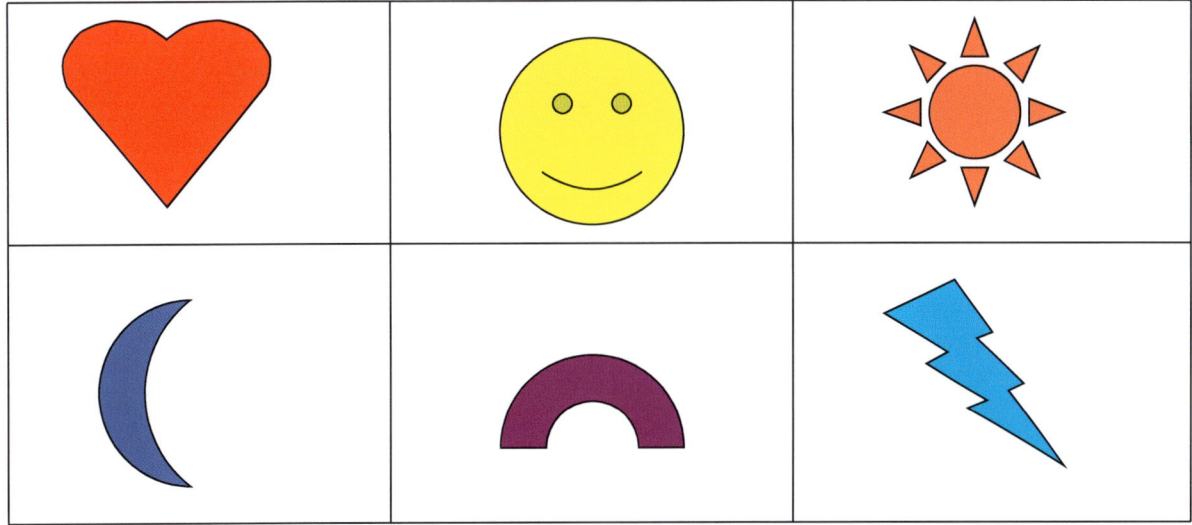

Come & Go Learn Chemistry with Me - Collect one Element AMU Card Set	Get enough of the elements H, Fe, and O to correctly make the following chemicals (figure out their formulas from the following names): **Iron (III) Hydroxide, Iron (III) Oxide, and Water.**	Get enough of the elements H, N, and O to correctly make the following chemicals (figure out their formulas from the following names): Nitric Acid (an oxyacid), Dinitrogen Pentoxide (a non-metal oxide), and Water (a molecule).	Get enough of the elements H, S, O, Na, and C to correctly make the following chemicals (figure out their formulas from the following names): Sodium Bicarbonate (salt of a weak acid), Sulfuric Acid (strong acid), Sodium Sulfate (salt), Carbon Dioxide (weak acid), Water.
Get enough of the elements H, C, and O to correctly make the following chemicals (figure out their formulas from the following names): Bicarbonate Hydrate (a hydrated compound), Bicarbonate (an anhydrous compound), Water, Carbon Monoxide, and Carbon Dioxide.			Get enough of the elements N, H, Cl, K, and O to correctly make the following chemicals (figure out their formulas from the following names): Ammonium Chloride (salt of a weak acid), Potassium Hydroxide (strong base), Ammonia, Water, and Potassium Chloride.
Get enough of the elements Cu, S, H, N, and O to correctly make the following chemicals (figure out their formulas from the following names): Copper (III) Sulfide, Nitric Acid, Copper (II) Nitrate, and Nitrous Oxide.			Get enough of the elements H, C, and O to correctly make the following chemicals (figure out their formulas from the following names): Bicarbonate Hydrate (a hydrated compound), Bicarbonate (an anhydrous compound), Water, Carbon Monoxide, and Carbon Dioxide.

(52) **Choose a banker. Select a Game Piece.**

(53) Each player **gets one full set of Element AMU Cards** at **the** start of the game **from** the **banker**. These are all of the elements used in the game (the entire set of elements in the periodic table).

(54) Every time a player **passes the "Come and Go Learn Chemistry with Me" box with his or her Game Piece,** he or she **gets another complete Element AMU set.**

(55) **To decide the order of play,** each player **selects a Step Card.** The **highest number of steps** on a Step Card **goes first, and so forth**.

(56) **On each player's turn, that player must select a Step Card** (the deck may be shuffled periodically – no pun intended).

(57) **Move one square for each of the steps shown on the Step Card along** the square of the Outer Path (**the colored squares**) with his or her **Game Piece**.

(58) After a player has moved, he or she can **decide whether or not to "claim" the square** on the Outer Path he or she landed on (provided no other player already has "claimed" it).

(59) If a player wants **to "claim" the square** on the Outer Path, he or she must **put at least one Element AMU Card** in the Inner Path square (**white square**) **underneath** the Outer Path (colored) square you "claimed." **Get a Claim Card from the banker the corresponding to the Outer Path Square (colored) that you wish to "claim."** The Element AMU Card you put in the Inner Path square must be one of the elements named in the Outer Path square you claimed.

(60) **Keep taking turns, moving around the Outer Path squares with your Game Piece after selecting a Step Card.**

(61) **If you land on an Outer Path square "claimed" by another player, then** you must **give the other player the same Element AMU Cards as the player has in the Inner Path square underneath it**. If you do not have the same Element AMU Cards, then you must give Element AMU Cards to the player greater than (more than) the AMU (atomic mass unit) of each element whose AMU card you do not have.

(62) **Keep taking turns, moving around the Outer Path squares with your Game Piece after selecting a Step Card.**

(63) **Next, work on getting all of the elements in each chemical compound's formula named in the Outer Path (colored) square you "claimed."** You must **put one Element AMU Card for each element that you need to correctly write the chemical formula of each named chemical.**

(64) **Keep taking turns, moving around the Outer Path squares** with your Game Piece after selecting a Step Card.

(65) **If you land on another player's Outer Path square, then** you must **give to the player "claiming" the square your Element AMU Cards corresponding to the player's Element AMU Cards**. If you do not have the same Element AMU Cards, then you must give Element AMU Cards to the player greater than (more than) the AMU (atomic mass unit) of each element whose Element AMU Card you do not have.

(66) **Keep taking turns.**

(67) **The person with the most Element AMU Cards when "time" is called wins the game.**

(68) **When a player runs out of AMU cards they must withdraw from the game.**

In this version of the periodic table of the elements you see the different types of the outermost atomic orbitals (s, p, d, f) that the "valence" electrons of the elements are found in. The negatively charged electrons "orbit" their atoms' small nuclei (where the positively charged protons and electrically neutral neutrons are). This gives the s-block, the p-block, the d-block, and the f-block elemnts.

s-block

#	Symbol	Name	Mass	Config
1	H	Hydrogen	1.01	$1s^1$
3	Li	Lithium	6.94	$2s^1$
4	Be	Beryllium	9.01	$2s^2$
11	Na	Sodium	22.99	$3s^1$
12	Mg	Magnesium	24.31	$3s^2$
19	K	Potassium	39.10	$4s^1$
20	Ca	Calcium	40.08	$4s^2$
37	Rb	Rubidium	85.469	$5s^1$
38	Sr	Strontium	87.82	$5s^2$
55	Cs	Cesium	132.91	$6s^1$
56	Ba	Barium	137.33	$6s^2$
87	Fr	Francium	(223)	$7s^1$
88	Ra	Radium	226.03	$7s^2$

d-block

#	Symbol	Name	Mass	Config
21	Sc	Scandium	44.96	$4s^23d^1$
22	Ti	Titanium	47.87	$4s^23d^2$
23	V	Vanadium	50.94	$4s^23d^3$
24	Cr	Chromium	52.00	$4s^13d^5$
25	Mn	Manganese	54.94	$4s^23d^5$
26	Fe	Iron	55.85	$4s^23d^6$
27	Co	Cobalt	58.93	$4s^23d^7$
28	Ni	Nickel	58.69	$4s^23d^8$
29	Cu	Copper	63.54	$4s^13d^{10}$
30	Zn	Zinc	65.39	$4s^23d^{10}$
39	Y	Yttrium	88.908	$5s^24d^1$
40	Zr	Zirconium	91.22	$5s^24d^2$
41	Nb	Nobelium	92.906	$5s^14d^4$
42	Mo	Molybdenum	95.94	$5s^14d^5$
43	Tc	Technetium	(97)	$5s^24d^5$
44	Ru	Rutheniu	102.91	$5s^14d^7$
45	Rh	Rhodium	102.91	$5s^14d^8$
46	Pd	Palladium	106.4	$4d^{10}$
47	Ag	Silver	107.87	$5s^14d^{10}$
48	Cd	Cadmium	112.4	$5s^24d^{10}$
71	Lu	Lutetium	174.97	$6s^25d^1$
72	Hf	Hafnium	172.49	$6s^25d^2$
73	Ta	Tantalum	190.95	$6s^25d^3$
74	W	Tungsten	183.85	$6s^25d^4$
75	Re	Rhenium	186.21	$6s^25d^5$
76	Os	Osmium	190.2	$6s^25d^6$
77	Ir	Iridium	192.22	$6s^25d^7$
78	Pt	Platinum	195.09	$6s^15d^9$
79	Au	Gold	196.97	$6s^15d^{10}$
80	Hg	Mercury	200.59	$6s^25d^{10}$
103	Lr	Lawrencium	(260)	$7s^26d^1$
104	Unq	Unnilquadium	(261)	$7s^26d^2$
105	Unp	Unnilpentium	(262)	$7s^26d^3$
106	Unh	Unnilhexium	(263)	$7s^26d^4$
107	Uns	Unnilseptium	(262)	$7s^26d^5$
108	Uno	Unniloctium	(265)	$7s^26d^6$
110	Uun	Ununilium	(269)	$7s^26d^6$
111	Uuu	Unununium	(272)	
112	Uub			

p-block

#	Symbol	Name	Mass	Config
5	B	Boron	10.81	$2s^22p^1$
6	C	Carbon	12.01	$2s^22p^2$
7	N	Nitrogen	14.01	$2s^22p^3$
8	O	Oxygen	16.00	$2s^22p^4$
9	F	Fluorine	19.00	$2s^22p^5$
10	Ne	Neon	20.18	$2s^22p^6$
13	Al	Aluminum	26.98	$3s^23p^1$
14	Si	Silicon	28.09	$3s^23p^2$
15	P	Phosphorus	30.97	$3s^23p^3$
16	S	Sulfur	32.07	$3s^23p^4$
17	Cl	Chlorine	35.45	$3s^23p^5$
18	Ar	Argon	39.90	$3s^23p^6$
31	Ga	Gallium	67.72	$4s^24p^1$
32	Ge	Germanium	72.61	$4s^24p^2$
33	As	Arsenic	74.92	$4s^24p^3$
34	Se	Selenium	78.96	$4s^24p^4$
35	Br	Bromine	79.90	$4s^24p^5$
36	Kr	Krypton	83.80	$4s^24p^6$
49	In	Indium	114.82	$5s^25p^1$
50	Sn	Tin	116.69	$5s^25p^2$
51	Sb	Antimony	121.75	$5s^25p^3$
52	Te	Tellurium	127.6	$5s^25p^4$
53	I	Iodine	126.9	$5s^25p^5$
54	Xe	Xenon	131.3	$5s^25p^6$
81	Tl	Thallium	204.37	$6s^26p^1$
82	Pb	Lead	207.2	$6s^26p^2$
83	Bi	Bismuth	208.96	$6s^26p^3$
84	Po	Polonium	(208)	$6s^26p^4$
85	At	Astatine	(210)	$6s^26p^5$
86	Rn	Radon	(222)	$6s^26p^6$
2	He	Helium	4.00	$1s^2$

f-block

#	Symbol	Name	Mass	Config
57	La	Lanthanum	138.91	$6s^25d^1$
58	Ce	Cesium	140.12	$6s^25d^14f^1$
59	Pr	Praseodymium	140.907	$6s^24f^3$
60	Nd	Neodymium	144.24	$6s^24f^4$
61	Pm	Promethium	(145)	$6s^24f^5$
62	Sm	Samarium	150.36	$6s^24f^6$
63	Eu	Europium	151.965	$6s^24f^7$
64	Gd	Gadolinium	157.25	$6s^25d^14f^7$
65	Tb	Terebium	158.93	$6s^24f^9$
66	Dy	Dysprosium	162.50	$6s^24f^{10}$
67	Ho	Holmium	164.93	$6s^24f^{11}$
68	Er	Erbium	167.26	$6s^24f^{12}$
69	Tm	Thulium	168.93	$6s^24f^{13}$
70	Yb	Ytterbium	173.04	$6s^24f^{14}$
89	Ac	Actinium	(227)	$7s^26d^1$
90	Th	Thallium	232.04	$7s^26d^2$
91	Pa	Protactinium	231.03	$7s^26d^15f^2$
92	U	Uranium	238.029	$7s^26d^15f^3$
93	Np	Neptunium	(237)	$7s^26d^15f^4$
94	Pu	Plutonium	(244)	$7s^25f^6$
95	Am	Americium	(243)	$7s^25f^7$
96	Cm	Curium	(247)	$7s^26d^15f^7$
97	Bk	Berkelium	(247)	$7s^25f^9$
98	Cf	Californium	(251)	$7s^25f^{10}$
99	Es	Einsteinium	(254)	$7s^25f^{11}$
100	Fm	Fermium	(257)	$7s^25f^{12}$
101	Md	Mendelevium	(258)	$7s^25f^{13}$
102	No	Nobelium	(259)	$7s^25f^{14}$

Move one (1) step.	Move two (2) steps.	Move three (3) steps.	Move four (4) steps.
Moving 1 step is to remind you of the element Hydrogen's one proton. Because Hydrogen has one proton, its atomic number is 1. The chemical symbol for the element Hydrogen is H. Because H has one proton and an atomic number = 1, it may be written as $_1H$.	Moving 2 steps is to remind you of the element Helium's two protons. Because Helium has two protons, its atomic number is 2. The chemical symbol for the element Helium is He. Because He has two protons and an atomic number = 2, it may be written as $_2He$.	Moving 3 steps is to remind you of the element Lithium's three protons. Because Lithium has three protons, its atomic number is 3. The chemical symbol for the element Lithium is Li. Because Li has three protons and an atomic number = 3, it may be written as $_3Li$.	Moving 4 steps is to remind you of the element Beryllium's four protons. Because Beryllium has four protons, its atomic number is 4. The chemical symbol for the element Beryllium is Be. Because Be has four protons and an atomic number = 4, it may be written as $_4Be$.
Move five (5) steps.	**Move six (6) steps.**	**Move seven (7) steps.**	**Move eight (8) steps.**
Moving 5 steps is to remind you of the element Boron's five protons. Because Boron has five protons, its atomic number is 5. The chemical symbol for the element Boron is B. Because B has five protons and an number = 5, it may be written as $_5B$.	Moving 6 steps is to remind you of the element Carbon's six protons. Because Carbon has six protons, its atomic number is 6. The chemical symbol for the element Carbon is C. Because C has six protons and an atomic number = 6, it may be written as $_6C$.	Moving 7 steps is to remind you of the element Nitrogen's seven protons. Because Nitrogen has seven protons, its atomic number is 7. The chemical symbol for the element Nitrogen is N. Because N has seven protons and an atomic number = 7, it may be written as $_7N$.	Moving 8 steps is to remind you of the element Oxygen's eight protons. Because Oxygen has eight protons, its atomic number is 8. The chemical symbol for the element Oxygen is O. Because O has eight protons and an atomic number = 8, it may be written as $_8O$.

Chemical Builder II (Part 4) Claim Cards (Answers too) ™ © 2006 GH Gebhart - Reproducible

Classroom Document

Get enough of the elements H, Fe and O to correctly make the following chemicals (figure out their formulas from the following names): **Iron (III) Hydroxide, Iron (III) Oxide, and Water. Answer: Fe(OH)$_3$, Fe$_2$O$_3$, and H$_2$O.**

Get enough of the elements H, N, and O to correctly make the following chemicals (figure out their formulas from the following names): Nitric Acid (an oxyacid), Dinitrogen Pentoxide (a non-metal oxide), and Water (a molecule). Answer: HNO$_3$, N$_2$O$_5$, and H$_2$O.

Get enough of the elements N, H, C, O, S, and Na to correctly make the following chemicals (figure out their formulas from the following names): Sodium Bicarbonate (salt of a weak acid), Sulfuric Acid (strong acid), Sodium Sulfate (salt), Carbon Dioxide (weak acid), Water. Answer: NaHCO$_3$, H$_2$SO$_4$, Na$_2$SO$_4$, CO$_2$, H$_2$O

Get enough of the elements H, C, and O to correctly make the following chemicals (figure out their formulas from the following names): Bicarbonate Hydrate (a hydrated compound), Bicarbonate (an anhydrous compound), Water, Carbon Monoxide, and Carbon Dioxide. Answer: H$_2$C$_2$O$_4$*2H$_2$O, H$_2$C$_2$O$_4$, H$_2$O, CO, CO$_2$

Get enough of the elements N, H, Cl, K, and O to correctly make the following chemicals (figure out their formulas from the following names): Ammonium Chloride (salt of a weak acid), Potassium Hydroxide (strong base), Ammonia, Water, and Potassium Chloride. Answer: NH$_4$Cl, KOH, NH$_3$, H$_2$O, KCl

Get enough of the elements Cu, S, H, N, and O to correctly make the following chemicals (figure out their formulas from the following names): Copper (III) Sulfide, Nitric Acid, Copper (II) Nitrate, and Nitrous Oxide. Answer: CuS, HNO$_3$, Cu(NO$_3$)$_2$, S$_8$, NO$_2$

Get enough of the elements H, C, and O to correctly make the following chemicals (figure out their formulas from the following names): Bicarborate Hydrate (a hydrated compound), Bicarbonate (an anhydrous compound), Water, , Carbon Monoxide, and Carbon Dioxide. H$_2$C$_2$O$_4$*2H$_2$O, H$_2$C$_2$O$_4$, H$_2$O, CO, CO$_2$

Get enough of the elements H, C, and O to correctly make the following chemicals (figure out their formulas from the following names): Bicarborate Hydrate (a hydrated compound), Bicarbonate

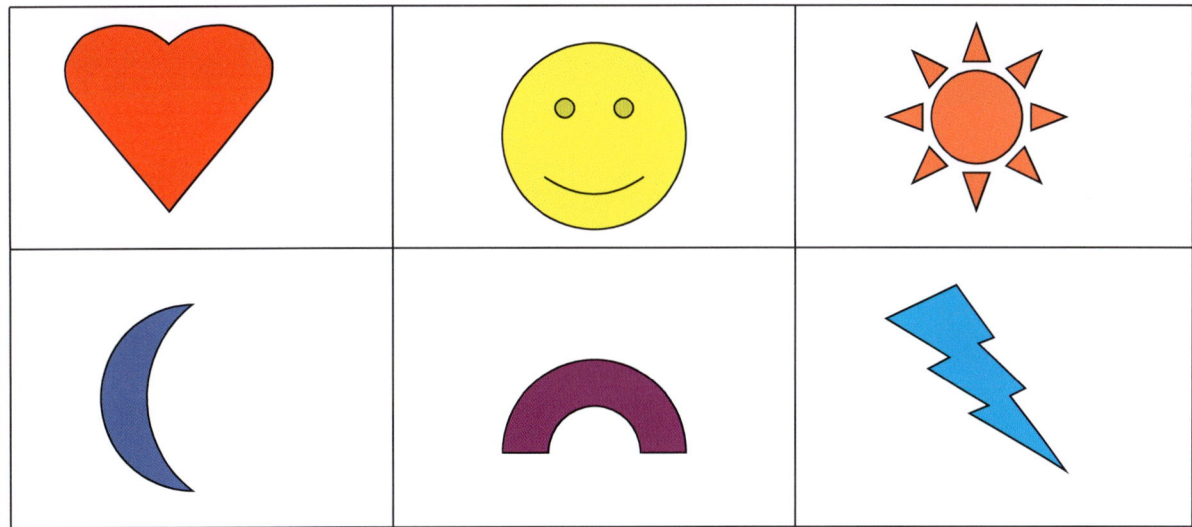

Come & Go Learn Chemistry with Me - Collect one Element AMU Card Set	Get enough of the elements H, C, and O to correctly make the following chemicals (figure out their formulas from the following names): Bicarbonate Hydrate (a hydrated compound), Bicarbonate (an anhydrous compound), and Water (a molecule).	Get enough of the elements Cu, S, O, and H to correctly make the following chemicals (figure out their formulas from the following names): Copper (II) Sulfate Hydrate, Copper (II) Sulfate, Water (Hydrated Material, Anhydrous Compound, and Water).	Get enough of the elements Al, Ni, S, and O to correctly make the following chemicals (figure out their formulas from the following names): Aluminum (metal), Nickel (III) Sulfate (ionic compound), Nickel (metal), Aluminum (II) Sulfate (ionic compound).
Get enough of the elements Cu, S, H, N, and O to correctly make the following chemicals (figure out their formulas from the following names): Copper (III) Sulfide, Nitric Acid, Copper (II) Nitrate, and Nitrous Oxide.			Get enough of the elements Cl, K, and I to correctly make the following chemicals (figure out their formulas from the following names): Chlorine (molecule), Potassium Iodide (Ionic compound), Potassium Chloride (ionic compound), Iodine (molecule)..
Get enough of the elements K, P, and O to correctly make the following chemicals (figure out their formulas from the following names): Potassium, TetraPhosphate HeptaOxygen, and Potassium Phosphate.			Get enough of the elements Na, H, and O to correctly make the following chemicals (figure them out from the following names): Sodium (metal), Water (molecule), Sodium Hydroxide (ionic compound), Hydrogen (molecule).

(69)   **Choose** a **banker. Select a Game Piece.**

(70)   Each player **gets one full set of Element AMU Cards** at **the** start of the game **from** the **banker.** These are all of the elements used in the game (the entire set of elements in the periodic table).

(71)   Every time a player **passes the "Come and Go Learn Chemistry with Me" box with his or her Game Piece,** he or she **gets another complete Element AMU set.**

(72)   **To decide the order of play,** each player **selects a Step Card.** The **highest number** of **steps** on a Step Card **goes first, and so forth.**

(73)   **On each player's turn, that player must select a Step Card** (the deck may be shuffled periodically – no pun intended).

(74)   **Move one square for each of the steps shown on the Step Card along** the square of the Outer Path (**the colored squares**) with his or her **Game Piece.**

(75)   After a player has moved, he or she can **decide whether or not to "claim" the square** on the Outer Path he or she landed on (provided no other player already has "claimed" it).

(76)   If a player wants **to "claim" the square** on the Outer Path, he or she must **put at least one Element AMU Card** in the Inner Path square (**white square**) **underneath** the Outer Path **(colored) square you "claimed." Get a Claim Card from the banker** the corresponding to the Outer Path Square (colored) that you wish to "claim." The Element AMU Card you put in the Inner Path square must be one of the elements named in the Outer Path square you claimed.

(77)   **Keep taking turns, moving around the Outer Path squares with your Game Piece after selecting a Step Card.**

(78)   **If you land on an Outer Path square "claimed" by another player, then** you must **give the other player the same Element AMU Cards as the player has in the Inner Path square underneath it.** If you do not have the same Element AMU Cards, then you must give Element AMU Cards to the player greater than (more than) the AMU (atomic mass unit) of each element whose AMU card you do not have.

(79)   **Keep taking turns, moving around the Outer Path squares with your Game Piece after selecting a Step Card.**

(80)   **Next, work on getting all of the elements in each chemical compound's formula named in the Outer Path (colored) square you "claimed." You must put one Element AMU Card for each element that you need to correctly write the chemical formula of each named chemical.**

(81)   **Keep taking turns, moving around the Outer Path squares** with your Game Piece after selecting a Step Card.

(82)   **If you land on another player's Outer Path square, then** you must **give to the player "claiming" the square your Element AMU Cards corresponding to the player's Element AMU Cards.** If you do not have the same Element AMU Cards, then you must give Element AMU Cards to the player greater than (more than) the AMU (atomic mass unit) of each element whose Element AMU Card you do not have.

(83)   **Keep taking turns.**

(84)   **The person with the most Element AMU Cards when "time" is called wins the game.**

(85)   **When a player runs out of AMU cards they must withdraw from the game.**

Chemical Builder II (Part 5) Element AMU Cards ™

© 2006 GH Gebhart - Reproducible Classroom Document

In this version of the periodic table of the elements you see the different types of the outermost atomic orbitals (s, p, d, f) that the "valence" electrons of the elements are found in. The negatively charged electrons "orbit" their atoms' small nuclei (where the positively charged protons and electrically neutral neutrons are). This gives the s-block, the p-block, the d-block, and the f-block elemnts.

1 **H** Hydrogen 1.01 $1s^1$																	**2** **He** Helium 4.00 $1s^2$
3 **Li** Lithium 6.94 $2s^1$	**4** **Be** Beryllium 9.01 $2s^2$											**5** **B** Boron 10.81 $2s^22p^1$	**6** **C** Carbon 12.01 $2s^22p^2$	**7** **N** Nitrogen 14.01 $2s^22p^3$	**8** **O** Oxygen 16.00 $2s^22p^4$	**9** **F** Fluorine 19.00 $2s^22p^5$	**10** **Ne** Neon 20.18 $2s^22p^6$
11 **Na** Sodium 22.99 $3s^1$	**12** **Mg** Magnesium 24.31 $3s^2$											**13** **Al** Aluminum 26.98 $3s^23p^1$	**14** **Si** Silicon 28.09 $3s^23p^2$	**15** **P** Phosphorus 30.97 $3s^23p^3$	**16** **S** Sulfur 32.07 $3s^23p^4$	**17** **Cl** Chlorine 35.45 $3s^23p^5$	**18** **Ar** Argon 83.90 $3s^23p^6$
19 **K** Potassium 39.10 $4s^1$	**20** **Ca** Calcium 40.08 $4s^2$	**21** **Sc** Scandium 44.96 $4s^23d^1$	**22** **Ti** Titanium 47.87 $4s^23d^2$	**23** **V** Vanadium 50.94 $4s^23d^3$	**24** **Cr** Chromium 52.00 $4s^13d^5$	**25** **Mn** Manganese 54.94 $4s^23d^5$	**26** **Fe** Iron 55.85 $4s^23d^6$	**27** **Co** Cobalt 58.93 $4s^23d^7$	**28** **Ni** Nickel 58.69 $4s^23d^8$	**29** **Cu** Copper 63.54 $4s^13d^{90}$	**30** **Zn** Zinc 65.39 $4s^23d^{10}$	**31** **Ga** Gallium 67.72 $4s^24p^1$	**32** **Ge** Germanium 72.61 $4s^24p^2$	**33** **As** Arsenic 74.92 $4s^24p^3$	**34** **Se** Selenium 78.96 $4s^24p^4$	**35** **Br** Bromine 79.90 $4s^24p^5$	**36** **Kr** Krypton 83.80 $4s^24p^6$
37 **Rb** Rubidium 85.469 $5s^1$	**38** **Sr** Strontium 87.82 $5s^2$	**39** **Y** Yttrium 88.908 $5s^24d^1$	**40** **Zr** Zirconium 91.22 $5s^24d^2$	**41** **Nb** Nobelium 92.906 $5s^14d^4$	**42** **Mo** Molybdenum 95.94 $5s^24d^4$	**43** **Tc** Technetium (97) $5s^24d^5$	**44** **Ru** Ruthenium 102.91 $5s^14d^7$	**45** **Rh** Rhodium 102.91 $5s^14d^8$	**46** **Pd** Palladium '106.4 $4d^{10}$	**47** **Ag** Silver 107.87 $5s^14d^{10}$	**48** **Cd** Cadmium 112.4 $5s^24d^{10}$	**49** **In** Indium 114.82 $5s^25p^1$	**50** **Sn** Tin 116.69 $5s^25p^2$	**51** **Sb** Antimony 121.75 $5s^25p^3$	**52** **Te** Tellurium 127.6 $5s^25p^4$	**53** **I** Iodine 126.9 $5s^25p^5$	**54** **Xe** Xenon 131.3 $5s^25p^6$
55 **Cs** Cesium 132.91 $6s^1$	**56** **Ba** Barium 137.33 $6s^2$	**71** **Lu** Lutetium 174.97 $6s^25d^1$	**72** **Hf** Hafnium 172.49 $6s^25d^2$	**73** **Ta** Tantalum 190.95 $6s^25d^3$	**74** **W** Tungston 183.85 $6s^25d^4$	**75** **Re** Rhenium 186.21 $6s^25d^5$	**76** **Os** Osmium 190.2 $6s^25d^6$	**77** **Ir** Iridium 192.22 $6s^25d^7$	**78** **Pt** Platinum 195.09 $6s^15d^9$	**79** **Au** Gold 196.97 $6s^15d^{10}$	**80** **Hg** Mercury 200.59 $6s^25d^{10}$	**81** **Tl** Thallium 204.37 $6s^26p^1$	**82** **Pb** Lead 207.2 $6s^26p^2$	**83** **Bi** Bismuth 208.96 $6s^26p^3$	**84** **Po** Polonium (208) $6s^26p^4$	**85** **At** Astatine (210) $6s^26p^5$	**86** **Rn** Radon (222) $6s^26p^6$
87 **Fr** Francium (223) $7s^1$	**88** **Ra** Radium 226.03 $7s^2$	**103** **Lr** Lawrencium (260) $7s^26d^1$	**104** **Unq** Unnilquadium (261) $7s^26d^2$	**105** **Unp** Unnilpentium (262) $7s^26d^3$	**106** **Unh** Unnilhexium (263) $7s^26d^4$	**107** **Uns** Unnilseptium (262) $7s^26d^5$	**108** **Uno** Unniloctium (265) $7s^26d^6$	**109** **Une** Unnilunium (269) $7s^26d^7$	**110** **Uun** Ununnilium (272)	**111** **Uuu** Unununium (272)	**112** **Uub** (272)						

57 **La** Lanthanum 138.91 $6s^25d^1$	**58** **Ce** Cesium 140.12 $6s^25d^14f^1$	**59** **Pr** Praseodymium 140.907 $6s^24f^3$	**60** **Nd** Neodymium 144.24 $6s^24f^4$	**61** **Pm** Promethium (145) $6s^24f^5$	**62** **Sm** Samarium 150.36 $6s^24f^6$	**63** **Eu** Europium 151.965 $7s^25f^6$	**64** **Gd** Gadolinium 157.25 $6s^25d^14f^7$	**65** **Tb** Terebium 158.93 $6s^24f^9$	**66** **Dy** Dysprosium 162.50 $7s^25f^{10}$	**67** **Ho** Holmium 164.93 $6s^24f^{11}$	**68** **Er** Erbium 167.26 $6s^24f^{12}$	**69** **Tm** Thulium 168.93 $6s^24f^{12}$	**70** **Yb** Ytterbium 173.04 $6s^24f^{14}$
89 **Ac** Actinium (227) $7s^26d^1$	**90** **Th** Thallium 232.04 $6s^26d^2$	**91** **Pa** Protactinium 231.03 $7s^26d^15f^2$	**92** **U** Uranium 238.029 $7s^26d^15f^3$	**93** **Np** Neptunium (237) $7s^26d^15f^4$	**94** **Pu** Plutonium (244) $7s^25f^6$	**95** **Am** Americium (243) $7s^25f^7$	**96** **Cm** Curium (247) $7s^26d^15f^7$	**97** **Bk** Berkelium (247) $7s^25f^9$	**98** **Cf** Californium (251) $7s^25f^{10}$	**99** **Es** Einsteinium (254) $7s^25f^{11}$	**100** **Fm** Fermium (257) $7s^25f^{12}$	**101** **Md** Mendelevium (258) $7s^25f^{13}$	**102** **No** Nobelium (259) $7s^25f^{14}$

Move one (1) step.	**Move two (2) steps.**	**Move three (3) steps.**	**Move four (4) steps.**

Move one (1) step.

Moving 1 step is to remind you of the element Hydrogen's one proton. Because Hydrogen has one proton, its atomic number is 1. The chemical symbol for the element Hydrogen is H. Because H has one proton and an atomic number = 1, it may be written as $_1H$.

Move two (2) steps.

Moving 2 steps is to remind you of the element Helium's two protons. Because Helium has two protons, its atomic number is 2. The chemical symbol for the element Helium is He. Because He has two protons and an atomic number = 2, it may be written as $_2He$.

Move three (3) steps.

Moving 3 steps is to remind you of the element Lithium's three protons. Because Lithium has three protons, its atomic number is 3. The chemical symbol for the element Lithium is Li. Because Li has three protons and an atomic number = 3, it may be written as $_3Li$.

Move four (4) steps.

Moving 4 steps is to remind you of the element Beryllium's four protons. Because Beryllium has four protons, its atomic number is 4. The chemical symbol for the element Beryllium is Be. Because Be has four protons and an atomic number = 4, it may be written as $_4Be$.

Move five (5) steps.

Moving 5 steps is to remind you of the element Boron's five protons. Because Boron has five protons, its atomic number is 5. The chemical symbol for the element Boron is B. Because B has five protons and an number = 5, it may be written as $_5B$.

Move six (6) steps.

Moving 6 steps is to remind you of the element Carbon's six protons. Because Carbon has six protons, its atomic number is 6. The chemical symbol for the element Carbon is C. Because C has six protons and an atomic number = 6, it may be written as $_6C$.

Move seven (7) steps.

Moving 7 steps is to remind you of the element Nitrogen's seven protons. Because Nitrogen has seven protons, its atomic number is 7. The chemical symbol for the element Nitrogen is N. Because N has seven protons and an atomic number = 7, it may be written as $_7N$.

Move eight (8) steps.

Moving 8 steps is to remind you of the element Oxygen's eight protons. Because Oxygen has eight protons, its atomic number is 8. The chemical symbol for the element Oxygen is O. Because O has eight protons and an atomic number = 8, it may be written as $_8O$.

Chemical Builder II (Part 5) Claim Cards (Answers too) ™ © 2006 GH Gebhart - Reproducible

Classroom Document

Get enough of the elements H, C, and O to correctly make the following chemicals (figure out their formulas from the following names): Bicarbonate Hydrate (a hydrated compound), Bicarbonate (an anhydrous compound), Water, Carbon Monoxide, and Carbon Dioxide. Answer: $H_2C_2O_4*2H_2O$, $H_2C_2O_4$, H_2O, CO, and CO_2

Get enough of the elements Cu, S, O, and H to correctly make the following chemicals (figure out their formulas from the following names): Copper (II) Sulfate Hydrate, Copper (II) Sulfate, Water (Hydrated Material, Anhydrous Compound, and Water). Answer: $CuSO_4*2H_2O$, $CuSO_4$, H_2O, CuO, and SO_3.

Get enough of the elements Al, Ni, S, and O to correctly make the following chemicals (figure out their formulas from the following names): Aluminum (metal), Nickel (III) Sulfate (ionic compound), Nickel (metal), Aluminum (II) Sulfate (ionic compound). Answer: Al, $NiSO_4$, Ni, and $Al(SO_4)_3$.

Get enough of the elements Cu, S, H, N, and O to correctly make the following chemicals (figure out their formulas from the following names): Copper (III) Sulfide, Nitric Acid, Copper (II) Nitrate, and Nitrous Oxide. Answer: CuS, HNO_3, $Cu(NO_3)_2$, S_8, and NO_2.

Get enough of the elements Cl, K, and I to correctly make the following chemicals (figure out their formulas from the following names): Chlorine (molecule), Potassium Iodide (Ionic compound), Potassium Chloride (ionic compound), Iodine (molecule). Answer: Cl_2, KI, KCl, and I_2.

Get enough of the elements Na, H, and O to correctly make the following chemicals (figure them out from the following names): Sodium (metal), Water (molecule), Sodium Hydroxide (ionic compound), Hydrogen (molecule). Answer: Na, H_2O, NaOH, and H_2.

Get enough of the elemerts K, P, and O to correctly make the following chemicals (figure out their formulas from the following names): Potassium TetraPhosphate HeptaOxygen, and Potassium Phosphate. Answer: K, P_4O_7, and K_3PO_3

Chemical Builder II (Part 5) Game Pieces ™ © 2006 GH Gebhart - Reproducible Classroom Document

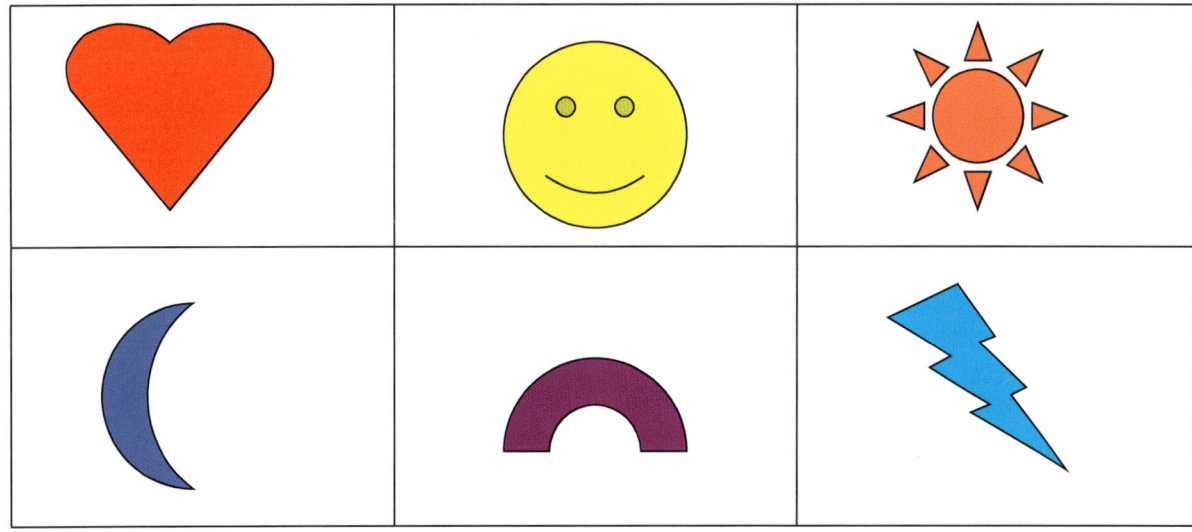

Printed in Great Britain
by Amazon